Archimedes Principle and the Law of Floatation

© sciencebod 2017

Preface

Archimedes principle and the Law of floatation both form the bedrock of understanding the interactions between a solid and a fluid in which it is immersed. These principles are among the most important in fundamental classical mechanics.

We understand it is very important for fresh science students to grasp the concepts and ideas put forward in these principles, and so we have taken time to present them in very lucid and concise manner. We urge students to be relaxed as they read through this book because the authors have presented the ideas in very appealing and interesting manner with focus on understanding, rather than just another science book.

There are lots of problems used as examples to illustrate the concepts, and many more exercises for students to try on their own.

© sciencebod, okodan2003@gmail.com; +2348136094616, +2348062111865

Archimedes Principle and the Law of Floatation

Equal volume of different substances

1

An equal volume of different substances have different masses or weights. For example 1m³ of iron weighs differently from 1m³ of lead. Also 1m³ of water has weight different from that of 1m³ of palm oil. This is due to the differences in the property of the substances known as density

What is Density?

2

The density of an object is how much mass a unit volume of the substance has.

Density is therefore defined as the mass per unit volume of the substance.

That is, Density = $\dfrac{mass}{volume}$ \hfill (1)

The unit of Density

3

Going by the formula in equation (1), the unit density is therefore the unit of mass divided by that of volume. Since the S.I unit of mass is kg and that of volume is m³, it means that the S.I unit of volume is **kgm⁻³**.

In the laboratory however, smaller units of density (e.g. gcm⁻³) may be used. Just for clarity, 1 gcm⁻³ is equal to 1000 kgm⁻³ as illustrated below:

$\therefore \dfrac{1g}{cm^3} = \dfrac{1 \times 10^{-3} kg}{10^{-6} m^3} = 10^3 \, kgm^{-3}$ or $1000 \, kgm^{-3}$ \hfill (This is because 1 g = 10^{-3} kg, and $1 \text{ cm}^3 = 10^{-6} \text{ m}^3$)

Understand what a density value means

4

If we say that the density of water is 1 gcm^{-3}, it means that 1 cm^3 of water has a mass of 1 g, and if we say that the density of water is 1000 kgm^{-3}, it means that 1 m^3 of water has a mass of 1000 kg.

Similarly, that the density of iron is 7900 kgm^{-3} simply means that 1 m^3 of iron has a mass of 7900 kg

Density values of some solids and liquids

5

The following table shows the density values of some solids and liquids
Table 1.

Material	Density kgm^{-3}
Aluminum	2.7×10^3
Copper	8.9×10^3
Gold	19.3×10^3
Glass	2.6×10^3
Lead	11.3×10^3
Platinum	21.5×10^3
Iron	7.9×10^3
Steel (variable)	7.8×10^3
Ice (at $0^0 c$)	0.92×10^3
Water (at $4^0 c$)	1×10^3
Mercury	13.6×10^3
Sand (variable)	2.6×10^3
Methylated spirit	0.8×10^3
Paraffin wax	0.9×10^3
Zinc	7.1×10^3
Turpentine	0.85×10^3
See water	1.03×10^3

The idea of Relative Density

6

If we say that the density of a particular liquid is 3 gcm^{-3}, it means that 1 cm^3 of that liquid has a mass of 3 g. This is 3 times that of water.

And so, relative to water, the liquid is 3 times denser.

The idea of relative density is to say how many times other substances are denser than water.

Water is chosen as a standard in this definition because water is the most common liquid.

To measure the Relative Density of a substance

7

It is intuitive from plan 6 that to measure the relative density of a substance, one just has to get the numerical mass (in g) of 1 cm^3 of the substance. This is correct but not the only way to go about it.

One could also measure out a given volume of that substance as well as the same volume of water (in this case, the volume mustn't be 1 cm^3, but whatever volume of the substance taken should be the same volume of water taken). And if we measure the mass of the substance and divide by the mass of the equal volume of water, this gives us the relative density of the substance.

By way of formula, the Relative density (Rd) is defined as:

$$Rd = \frac{mass\ or\ weight\ of\ a\ substance}{mass\ or\ weight\ of\ equal\ volume\ of\ water} \quad (2)$$

$$OR\ Rd = \frac{density\ of\ substance}{density\ of\ water} \quad (3)$$

And so, the density of a substance = Relative density of the substance × density of water.

Relative density is a dimensionless quantity

Note that since the relative density is a division of two quantities that have the same units (e.g. masses or weights or densities), the units cancel out and so relative density has no units.

We therefore say that the relative density of a substance is 5, to mean that the substance is 5 times denser than water. The density of this substance will therefore be 5 gcm^{-3} or 5×10^3 kgm^{-3}.

Consequently, the relative density of mercury is 13.6, but its density is 13.6 gcm^{-3} or 13.6×10^3 kgm^{-3}.

Having understood those, let's now attempt some questions.

Question 1 (JAMB 2006)

The relative density of Gold is 19.2, the volume of 2.4 kg of gold is
(A) $8 \times 10^{-3} m^3$ (B) $1.25 \times 10^{-4} m^3$ (C) $1.92 \times 10^{-4} m^3$ (D) $4.6 \times 10^{-3} m^3$

[Density of water = 10^3 kgm^{-3}]

Try to do it before looking at the solution in this plan

Solution:

Relative density of Gold = $\dfrac{density\ of\ Gold}{density\ of\ water}$

∴ $density\ of\ Gold = Relative\ density\ of\ Gold \times density\ of\ water$
$= 19.2 \times 10^3$ kgm^{-3}

Now $density = \dfrac{mass}{volume}$

Therefore volume = $\frac{mass}{density}$ = $\frac{2.4\ kg}{19.2 \times 10^3\ kgm^{-3}}$ = $1.25 \times 10^{-4}\ m^3$

So option B is correct!

Question 2 (JAMB 2005)

A $3m^3$ volume of liquid W of density 200 kgm^{-3} is mixed with another liquid L of volume $7m^3$ and density 150kgm^{-3}. The density of the mixture is
(A) 165 kgm^{-3} (B) 100 kgm^{-3} (C) 256 kgm^{-3} (D) 350 kgm^{-3}

Solution!

To get the density of the mixture, we need to get the total mass of the mixture and then divide it by the total volume of the mixture.

Mass of liquid W = density of W × volume of W
= 200 kgm^{-3} × 3 m^3 = 600 kg

Mass of liquid L = density of L × volume of L
= 150 kgm^{-3} × 7 m^3 = 1050 kg

Therefore, total mass of the two liquids = 600 kg + 1050 kg = 1650 kg
And total volume of the two liquids = (3 + 7) m^3 = 10 m^3

The density of the mixture is therefore = $\frac{total\ mass}{total\ volume}$ = $\frac{1650\ kg}{10\ m^3}$ = $165\ kgm^{-3}$

Option A is correct!

Question 3 (JAMB 2010)

The density of a certain oil on frying becomes 0.4 kgm^{-3} with a volume of 20 m^3. What will be its initial volume when its initial density is 0.8 kgm^{-3} assuming no loss of oil due to spillage.
(A) 12 m^3 (B) 10 m^3 (C) 8 m^3 (D) 5 m^3

Solution!

Density of oil on frying = 0.4 kgm^{-3}
Volume of oil on frying = 20 m^3
Therefore, Mass of oil on frying = Density of oil on frying × Volume of oil on frying
= 0.4 kgm^{-3} × 20 m^3 = 8 kg

Now, since there is no spillage,
Initial mass of oil = mass of oil on frying = 8 kg

$$\therefore initial\ volume\ of\ oil = \frac{initial\ mass\ of\ oil}{initial\ density\ of\ oil} = \frac{8\ kg}{0.8\ \text{kgm}^{-3}} = 10\ m^3$$

Option B is correct!

Question 4 (WAEC 2004)

Calculate the change in volume when 90 g of ice is completely melted.
[density of water = 1 gcm^{-3}, density of ice = 0.9 gcm^{-3}]

(A) 0.00 cm^3 (B) 9.00 cm^3 (C) 10.00 cm^3 (D) 90.00 cm^3

Solution!

Volume of 90g of ice = $\dfrac{Mass\ of\ ice}{density\ of\ ice}$ = $\dfrac{90\ g}{0.9\ gcm^{-3}}$ = 100 cm^3

On melting, the mass of the water does not change

∴ Volume of water = $\dfrac{Mass\ of\ water}{density\ of\ water}$ = $\dfrac{90\ g}{1\ gcm^{-3}}$ = 90 cm^3

And so, the change in volume = Volume of ice – Volume of water
= (100 - 90) cm³   = 10 cm³

Option C is correct!

Determining the volume of liquids and regularly-shaped solids

17

It is quite an easy and straight-forward task to determine the volume of a liquid; just pour the liquid into a measuring cylinder (or other measuring vessels), and you can read off its volume

It is also easy to measure the volume of regularly-shaped solids:
If it is a cube, just measure the length of one side (say x), then the volume of the cube is x^3.
If it is a cuboid, measure the length (l), the breadth (b), and the height (h), then the volume of the cuboid is l×b×h
If it is a sphere, just measure the radius (r), and its volume will be $\frac{4}{3} \times \pi \times r^3$
If it is a cylinder, measure the base radius (r), and the height (h), then the volume is π×r²×h
And so on,….

Now, what if the solid has a very irregular shape that we do not have a defined formula for it? E.g. the stones below!

Determining the volume of irregularly-shaped solids

18

A novel way to determine the volume of an irregularly-shaped solid is to drop it inside a liquid in which it does not dissolve or react. The volume of the liquid it displaces becomes the volume of the solid.

For example, to determine the volume of one of the irregularly-shaped stones in plan 17, we first put some water in a measuring cylinder as illustrated below. We read off the volume of the water, then gently drop the stone into it.

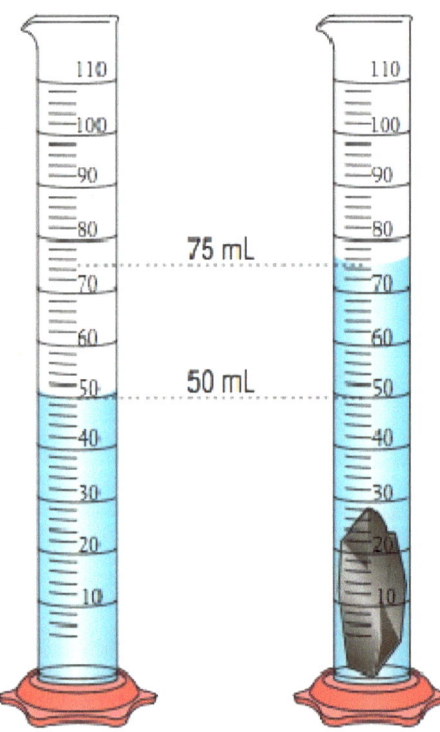

As the stone sinks into the water, it displaces some of the water, which is observed as a rise in the water level. The difference in the water levels before and after the stone was dropped is the volume of the water displaced, and this is equal to the volume of the irregular-shaped solid.

For the case of the illustration in the diagram above, the water levels before and after the stone was dropped are respectively 50 mL and 75 mL. The volume of the irregular-shaped stone is therefore 75 mL – 50 mL = 25 mL.

What really happens when solids are dropped in liquids?

19

As with the illustration in plan 18 above, anything a solid is dropped into a liquid in which it does not dissolve or reacted, the following happens:
1. The solid displaces a volume of the liquid which is equal to the volume of the solid immersed in the liquid
2. The solid losses some weight (also called upthrust)
3. The amount of weight which the solid losses is equal to the weight of the liquid it displaces.

These are the formulations of an important principle in this area of study known as the Archimedes principle, named after the Greek scientist that discovered it.

Archimedes Principle

20

The Archimedes principle states that: when a body is totally or partially immersed in a fluid (liquid or gas), it experiences an upthust which is equal to the weight of fluid displaced.

If we have clearly understood the principle, then we should be able to tackle the following questions.

Question 1 (JAMB 1997)

21

A cube of sides 0.1m hangs freely from a string. What is the upthrust on the cube when totally immersed in water?
(A) 1000 N   (B) 700 N   (C) 110 N   (D) 10 N
(Density of water is 1000 kgm^{-3}, g = 10 ms^{-2})

Solution

22

The volume of the cube = 0.1 m × 0.1 m × 0.1 m = 0.001 m³

If the cube is totally immersed in water, this is the same volume of water it will displace (0.001 m³).

Now, this volume of water has a mass = density of water × volume of water
= 1000 kgm⁻³ × 0.001 m³ = 1 kg

And therefore its weight = mass × g
= 1 kg × 10 ms⁻² = 10 N

Option D is correct

Question 2 (JAMB 1993)

23

An object of mass 400 g and density 600 kgm⁻³ is suspended with a string so that half of it is immersed in paraffin of density 900 kgm⁻³. The tension in the string is
(A) 1 N   (B) 3 N   (C) 4 N   (D) 5 N
(g = 10 ms⁻²)

Solution

24

The tension in the string is the weight of the object as it hangs on the string, and immersed in paraffin.

The weight of the object before it is immersed = mass × g
= 0.4 kg × 10 ms⁻² = 4 N

Now, when half of it is immersed in paraffin, the volume of paraffin displaced is half of the object's volume

$= \dfrac{1}{2} \times \dfrac{mass\ of\ object}{density\ of\ object} = \dfrac{1}{2} \times \dfrac{0.4\ kg}{600\ kgm^{-3}} = 3.3 \times 10^{-4}\ m^3$

This volume of paraffin displaced has a mass that is
= density of paraffin × volume of paraffin
= 900 kgm^{-3} × 3.3×10^{-4} m^3 = 0.3 kg

And so the weight of paraffin displaced = mass × g
= 0.3 kg × 10 ms^{-2} = 3 N

That is the value of the upthrust

And so the weight of the object during the immersion
= its weight before immersion – upthrust = (4 – 3) N = 1 N

That is the value of the tension in the string. And so option A is correct.

Floating and Sinking

25

Some objects float in water while others sink in it!
What are the conditions for an object to float or sink in water, or in any other liquid?

From Archimedes principle, it is obvious that an object that is immersed in a liquid will lose the same amount of weight (called upthrust) as the weight of the liquid it displaces.

The objects that float are the ones that displace their own weight of the liquid before they are completely immersed. This is the scenario:

I drop an object in water. As soon as the object starts going into the water (before it is completely immersed), it already starts displacing some amount of water. The more the object gets into water, the more the water it displaces. If, before it is completely immersed, it displaces as much water that has the same weight as itself, then the object hangs there. It stops going down because the upthrust has become equal to its own weight (it has lost all of its weight, and its weight in that water becomes zero). As the object hangs there, it is said to float. This is when and why objects float.

Some objects displace their own weight of the liquid even before they get half-immersed. Such objects will float with less than half of their body inside the liquid. Other objects displace their own weight only after more than half of their body is immersed, such objects will float with more than half of their body immersed in water.

On the other hand, objects that have not displaced their own weight of the liquid, even when they are completely immersed, will keep going down because they still have some weight in the liquid. This is when and why objects sink.

Big-sized objects with small masses usually float!

26

A big-sized object (that is, one with a large volume) will displace more volume of water as it goes down in water. If the mass of the object is small, then it is likely to have displaced its own weight of water before it is completely immersed. The object will therefore float. Such objects that have big volumes and small masses have very low densities.

On the other hand, an object that has small volume and much mass will likely not be able to displace enough volume of water that will have much weight as itself, so it will sink. Such objects with small volumes and large masses have very high densities.

In summary,... objects that are more dense than the liquids in which they are dropped will sink, while those that are less dense than the liquids in which they are dropped will float!

The Law of Floatation

27

The law of floatation states that: an object will float in a fluid (liquid or gas) when the upthrust exerted upon it by the fluid in which it floats equals the weight of the object.

In other words, a floating body displaces its own weight of the liquid in which it floats.

Factors affecting floatation

28

As already mentioned in plan 26, the density of an object is the major determinant as to whether or not the object will float in a liquid:
A body will float in a liquid if its density is less than that of the liquid.

So why do ships float in water?

29

Ships are made from materials that are several times denser than water. If only objects that are less dense than water can float in water, why do ships (which are made up of more dense materials) float in water?

An illustration to answer the question:
Consider the 2 objects (A and B) in the diagram below. They are both made up of the same metal material, but shaped differently; A is a solid sphere while B is shaped like a boat/ship. By the way object A is shaped, its density is directly the density of the metal used to make it, it will therefore sink in water since the density of this material is greater than that of water.

The situation is however different for object B. Ships are shaped in such a way (as object B) that their compositions do not only include the materials used to make them, but also of a vast amount of the atmosphere above them. The ship is therefore seen to be composed of the material as well as this vast amount of atmosphere as illustrated in the diagram below.

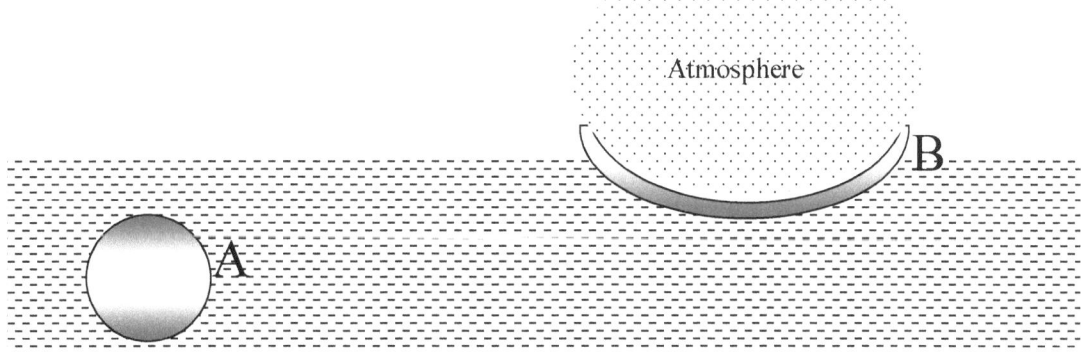

The ship therefore has a much larger volume than just that of the metal used to make it. In effect, the density of the ship is not just that of the metal material, but a combination of the material as well as the vast volume of the involved atmosphere. Since the density of air is relatively much lower than that of water, the density of the atmosphere-material combination is usually lower than that of water, and so ships float in water.

Summarily, ships float in water because they are shaped in a way that extends their volumes to include the vast portion of the atmosphere above them, and so their effective densities become less than that of water.

Thus for a body to float, its shape and the density of its material are important factors to consider.

Again!

A ship floats in water because its large volume displaces a large volume of water whose weight counterbalances the weight of the ship.

Similarly, a balloon filled with gas lighter than air will float if the weight of the balloon and content equals the upthrust of air on the balloon. This is one principle harnessed by scientists in sending balloons (carrying scientific equipment) to very high altitudes in the atmosphere.

Now it's time to test our understanding of the ideas treated so far!

Question 1 (JAMB 1992)

If a plastic sphere floats in water (density = $1000 kgm^{-3}$) with 0.5 of its volume submerged, and floats in oil with 0.4 of its volume submerged. The density of the oil is

(A) $800\ kgm^{-3}$ (B) $1200\ kgm^{-3}$ (C) $1250\ kgm^{-3}$ (D) $2000\ kgm^{-3}$

Solution!

32

The law of floatation tells us that the plastic sphere displaces its own mass of any liquid it floats on. That is:

Mass of water displaced when plastic sphere floats in water is equal to the mass of oil displaced when plastic sphere floats in oil
That is, mass of water displaced = mass of oil displaced

Since mass = density × volume
⇨ density of water × volume of water displaced = density of oil × volume of oil displaced

And so, density of oil = $\frac{density\ of\ water \times volume\ of\ water\ displaced}{volume\ of\ oil\ displaced}$

$= \frac{1000\ kgm^{-3} \times 0.5V}{0.4V}$ (where V is the volume of the plastic sphere)

$= 1250$ kgm^{-3}
And so option C is correct!

Question 2 (JAMB 2002)

33

A copper cube weighs 0.25 N in air, 0.17 N when completely immersed in paraffin oil, and 0.15 N when completely immersed in water. The ratio of upthrust in oil to upthrust in water is
(A) 3:5   (B) 4:5   (C) 7:10   (D) 13:10

Solution 34

The upthrust in oil is the loss in weight when the cube is immersed in oil
= 0.25 N – 0.17 N = 0.08 N

Similarly, the upthrust in water is = 0.25 N – 0.15 N = 0.10 N

Therefore, the ratio of the upthrust in oil to the upthrust in water is
0.08 : 0.10
= 4:5
Option B is correct!

Question 3 (JAMB 1999) 35

A solid weighs 10.0 N in air, 6.0 N when fully immersed in water and 7.0 N when fully immersed in a certain liquid X. Calculate the relative density of the liquid.
(A) $\frac{5}{3}$ (B) $\frac{4}{3}$ (C) $\frac{3}{4}$ (D) $\frac{7}{10}$

Solution 36

From equation (2):
Relative density of a substance = $\dfrac{mass\ of\ substance}{mass\ of\ an\ equal\ volume\ of\ water}$

Now, since the solid will displace the same volume of water as liquid X when immersed in them,

⇨ Relative density of liquid X = $\dfrac{mass\ of\ liquid\ X\ displaced}{mass\ of\ water\ displaced}$

= $\dfrac{upthrust\ in\ liquid\ X}{upthrust\ in\ water}$ = $\dfrac{10-7}{10-6}$ = $\frac{3}{4}$

Option C is correct.

Question 4 (JAMB 1998)

A solid of weight 0.6 N is totally immersed in oil and water respectively. If the upthrust in oil is 0.21 N and the relative density of oil is 0.875. Find the upthrust in water.

(A) 0.6 N (B) 0.36 N (C) 0.24 N (D) 0.18 N

Solution

Relative density of oil = $\dfrac{\text{upthrust in oil}}{\text{upthrust in water}}$

∴ upthrust in water = $\dfrac{\text{upthrust in oil}}{\text{relative denstiy of oil}}$

$= \dfrac{0.21\,N}{0.875} = 0.24\,N$

Option C is correct

Question 5 (JAMB 1994)

The mass of a specific gravity bottle is 15.2 g when it is empty. It is 24.8 g when filled with kerosene and 27.2 g when filled with distilled water. Calculate the relative density of kerosene.

(A) 1.25 (B) 1.10 (C) 0.90 (D) 0.80

Solution

40

Relative density of kerosene = $\dfrac{\text{mass of kerosene}}{\text{mass of an equal volume of water}}$

Now, when filled, the specific gravity bottle carries the same volume of both liquids

⇒ Relative density of kerosene = $\dfrac{\text{mass of kerosene in filled specific gravity bottle}}{\text{mass of water in filled specific gravity bottle}}$

$$= \dfrac{24.8 - 15.2}{27.2 - 15.2} = \dfrac{9.6}{12} = 0.8$$

Option D is right!

Hydrometers

41

The hydrometer is an instrument used to measure the relative density of liquids. It consists of a hollow narrow glass tube (called the stan) carried on a wide tube (called the bulb). It has a loaded end containing lead shots which enables the instrument to stand upright in liquids. The hydrometer is illustrated in the diagram below.

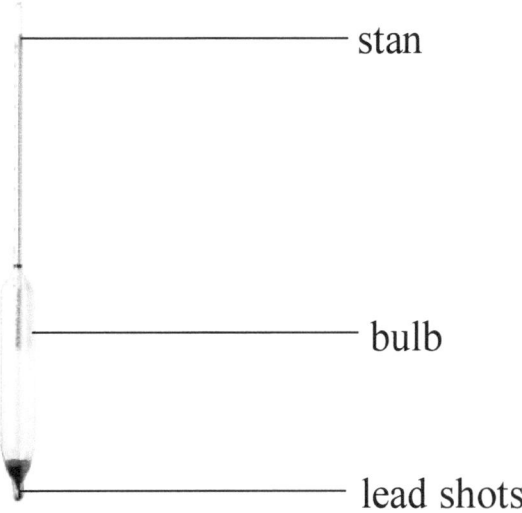

The wide bulb is to ensure that the upthrust of any liquid on the hydrometer is large enough to support its weight. The narrow Stan ensures that the instrument is very sensitive to small changes in density. The end loaded with lead shots is to ensure that the hydrometer floats upright and stable in liquids.

To measure the relative density of a liquid, the hydrometer is just dropped into the liquid.

The hydrometer is an Archimedes principle instrument!

The hydrometer functions based on the Archimedes principle. That is, the hydrometer will sink more in a liquid of low relative density than it does in a liquid of high relative density.

The Stan of the hydrometer is calibrated in such a way that the relative density of a liquid can be read directly. The lowest values of relative density are above and the highest values are below as shown in the diagram below. This is obviously because it will sink more in a liquid of low relative density, so the lower relative density marks should be on top.

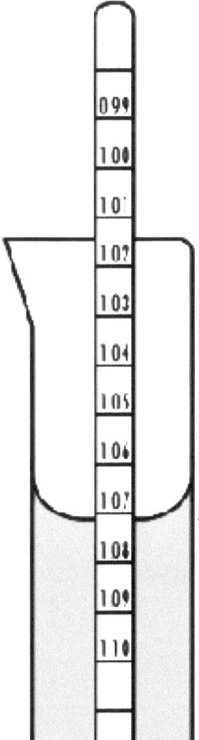

Put more scientifically; the length of the hydrometer Stan immersed in a liquid is inversely proportional to the relative density of the liquid.

Some practical uses of hydrometers

43

Hydrometers are used to test the concentration or relative density of acids in batteries.
They are used in testing the purity of liquids whose densities are known
They are also used in testing the quality of milk.

And that is where we draw the curtain for this topic on Archimedes principle and the law of floatation. But before we go, just a quick question in the next plan, and some exercises in plan 46 for you to test your understanding of the topic.

Question!

44

A hydrometer of mass 3.6 kg and volume 6.0×10^{-5} m³ floats in a liquid with $\frac{1}{5}$ of its volume above the liquid. Calculate the density of the liquid.

Solution!

45

Since the hydrometer floats with $\frac{1}{5}$ of its volume above the liquid, it means that $\frac{4}{5}$ of its volume is immersed in the liquid.

The volume of hydrometer immersed in liquid is therefore = $\left(\frac{4}{5} \times 6.0 \times 10^{-5}\right) m^3$

That is also the volume of the liquid displaced!

Now, we also know from the law of floatation that for the hydrometer to float, it must have displaced its own mass of the liquid it floats. It means that the mass o the liquid displaced is also 3.6 kg.

⇨ The volume of the liquid $\left(\frac{4}{5}\times 6.0\times 10^{-5}\right) m^3$ has a mass of 3.6 kg, and so the density of the liquid is:

$$\text{density} = \frac{mass}{volume} = \frac{3.6\ kg}{\frac{4}{5}\times 6.0\times 10^{-5}\ m^3} = 7.5\times 10^4\ \text{kgm}^{-3}$$

Exercises

(1) The mass of a stone is 15.0 g when completely immersed in water and 10.0 g when completely immersed in a liquid of relative density 2.0. The mass of the stone in air is
(A) 5.0 g   (B) 12.0 g   (C) 20.0 g   (D) 25.0 g

(2) An empty 60 liter petrol tank has a mass of 10 kg. Its mass when full of fuel of relative density 0.72 is
(A) 7.2 kg   (B) 33.2 kg   (C) 43.2 kg   (D) 53.2 kg
[Hint: 1 liter = 10^{-3} m^3]

(3) If a solid X floats in liquid P of relative density 2.0 and in liquid Q of relative density 1.5, it can be inferred that the
(A) Weight of P displaced is greater than that of Q
(B) Weight of P displaced is less than that of Q
(C) Volume of P displaced is greater than that of Q
(D) Volume of P displaced is less than that of Q

(4) The change in volume when 450 kg of ice is completely melted is
(A) 0.50 m^3 (B) 0.45 m^3 (C) 0.05 m^3 (D) 4.50 m^3
[density of ice = 900 kgm^{-3}, density of water = 10^3 kgm^{-3}]

(5) A liquid of volume 2.0 m^3 and density 1.00×10^3 kgm^{-3} is mixed with 3.00 m^3 of another liquid of density 0.90×10^3 kgm^{-3}. Find the density of the mixture.
[Assume there is no chemical reaction]
(A) 9.40×10^2 kgm^{-3} (B) 8.80×10^2 kgm^{-3} (C) 1.13×10^3 kgm^{-3} (D) 5.2×10^3 kgm^{-3}

(6) The apparent weight of a body fully immersed in water is 32 N and its weight in air is 96 N. Calculate the volume of the body [Density of water = 1000 kgm^{-3}, g = 10 ms^{-2}]
(A) 8.9×10^{-3} m^3 (B) 6.4×10^{-3} m^3 (C) 3.2×10^{-3} m^3 (D) 3.0×10^{-3} m^3

(7) A solid weighs 45 N and 15 N respectively in air and water, determine the relative density of the solid.
(A) 0.33   (B) 0.50   (C) 1.50   (D) 3.00

(8) A piece of metal of relative density 5.0 weighs 60 N in air. Calculate its weight when fully immersed in water.
(A) 4 N   (B) 5 N   (C) 48 N   (D) 60 N

(9) An object weighs 60.0 N in air, 48.2 N in a certain liquid L and 44.9 N in water. Calculate the relative density of liquid L.
(A) 3.300   (B) 1.279   (C) 0.932   (D) 0.782

(10) What is the density of a fuel of relative density 0.72 [density of water = 10^3 kgm^{-3}]
(A) 72 kgm^{-3} (B) 720 kgm^{-3} (C) 7200 kgm^{-3} (D) 72000 kgm^{-3}

(11) An empty density bottle weighs 2N. If it weighs 5N when filled with water, and 4N when filled with olive oil, the relative density of olive oil is
(A) $\frac{1}{3}$ (B) $\frac{2}{3}$ (C) $\frac{1}{5}$ (D) $\frac{2}{5}$

(12) An object of weight 10 N immersed in a liquid displaces a quantity of the liquid. If the liquid displaced weighs 6 N, determine the upthrust on the object.
(A) 20 N   (B) 10 N   (C) 6 N   (D) 4 N

(13) A block of volume 3×10^{-5} m^3 and density 2.5×10^3 kgm^{-3} is suspended from a spring balance with 2/3 of its volume immersed in a liquid of density 900 kgm^{-3}. Determine the reading of the spring balance [g = 10 ms^{-2}].
(A) 0.18 N   (B) 0.57 N   (C) 0.75 N   (D) 0.93 N

(14) A body of volume 0.046 m^3 is immersed in a liquid of density 980 kgm^{-3} with ¾ of its volume submerged. Calculate the upthrust on the body (g = 10 ms^{-2})
(A) 11.27 N   (B) 33.81 N   (C) 112.70 N   (D) 338.10 N

(15) A uniform cylindrical hydrometer of mass 20 g and cross-sectional area 0.54 cm² floats upright in a liquid. If 25 cm of its length is submerged, calculate the relative density of the liquid [Density of water = $1 gcm^{-3}$].
(A) 1.54   (B) 1.48   (C) 1.25   (D) 0.80

(16) A solid body will float in a liquid if its
(A) density is less than that of the liquid
(B) mass is equal to that of the liquid
(C) density is greater than that of the liquid
(D) mass is less than that of the liquid

(17) Which of the following statements about Archimedes principle is correct? The upthrust on a body is equal to the
(A) mass of fluid displaced
(B) weight of the body
(C) volume of the body
(D) weight of the fluid displaced

(18) A uniform solid cube material 10 cm on each side of mass 700 g is submerged in water. Which of the following best describes the behavior of the cube in water?
(A) the cube will melt after a period of time
(B) the cube will float in water
(C) the cube will sink in water
(D) the cube will rest at an equilibrium position in water

(19) A plastic sphere floats in water with 50% of its volume submerged. If it floats in glycerine with 40% of its volume submerged, the density of the glycerine is
(A) 1400 kgm^{-3} (B) 1250 kgm^{-3} (C) 500 kgm^{-3} (D) 1000 kgm^{-3}
[density of water = 10^3 kgm^{-3}]

(20) A hydrometer is an instrument used in measuring
(A) vapor pressure of a fluid
(B) density of a liquid
(C) relative humidity of a liquid
(D) relative density of a liquid

Solutions to Exercises

47

1. C
2. C
3. D
4. C
5. A
6. B
7. C
8. C
9. D
10. B
11. B
12. C
13. B
14. D
15. B
16. A
17. D
18. B
19. B
20. D

www.ingramcontent.com/pod-product-compliance
Lightning Source LLC
Chambersburg PA
CBHW040057250526

45473CB00043B/1849